ON LIVING WITH PETS

Approaching Your Animals with Empathy
and Living Together in Harmony

ROB ERTEMAN, DVM

Inspired by the teachings of Niek Brouw, MD

Copyright © 2024 Rob Erteman, Author
All rights reserved.

ISBN 978-1-304-21278-8

Table of Contents

Foreword	1
Relationship Between People and Animals in Freedom and in Health	3
Emotional Relationship with Our Animals	7
The Being Is the Purest Part of an Animal	11
Animals Are Emotional and Feel Life	13
Shock, Hunger, and Envy	16
Who Are We to Our Pets, and Who Are Our Pets?	18
Mind, Soul, and Spirit	20
Balance and Stability: Working with the Muscle Chains in Dogs	23
Dependence and Independence	28
Perception and Relationship	30
How Shock Manifests in Our Animals	32
Equality and Relativity: Feeling Our Pets and Ourselves and Understanding Each Other	35
Touching Pets to Change Their Consciousness	37
Accessing the Unknown in Pets	39
What Your Animals Need and What Your Animals Want. What Do You Give?	40
Perception Training and The Soul Body: Solving the "Unsolvable"	43

What is the Relationship Between Your Relationship with Your Animal and Your Animal's Health?	45
Do Cats Need Relationship with People?	48
Self-Awareness and Our Relationships with Dogs: What We Meet in Ourselves When Our Dogs Meet Their Shocks	50
Forcing Our Dogs To Do What We Think Is Best for Them (or Us)	52
Healing Developmental Gaps as an Alternative to Training: Instead of Suppression, Freedom to Feel and Move	54
Treating Aggressive Dogs Through Healing	56
Anxiety, Reactivity, and Aggression in Dogs: Finding Balance and Peace	58
Sharing Life with Our Pets: Achieving Harmony Without Control	61
Space and the Relationship with the Self	63
Living in Peace with Our Pets	65
Niek Brouw and the Foundation for Motor Development	67
Afterword by Anne Lamott	68

Foreword

I worked as a small animal veterinarian for 35 years. I was attracted to veterinary medicine because of the variety and freedom that the profession had to offer in the seventies, eighties, and nineties. People were very open emotionally with their pets and seemed to feel safe to expose and share their feelings with me. The quality of contact with people was rich and rewarding. As I grew as a veterinarian, really as a human being, I started to perceive and feel a lot of things about how deeply an animal could touch a person, the immense influence that people have over their pets, and the lengths that people would go to help their pets. It was not uncommon for me to hear that the loss of a pet felt more devastating than the loss of a human family member.

After seven years in practice, I realized that I didn't know what I was doing. Basic medical treatments solved immediate circumstances or crises, but I could feel that the underlying reasons for the circumstances were most often not addressed and that the animals were left with the underlying shock that produced the illness or the circumstance.

I left a thriving practice and went to a career counselor who suggested that I meet Dr. Niek Brouw. A profound diagnostician and therapist, Niek was known for his ability to see deeply and accurately into human beings and identify the developmental origins of disease. This changed my life. Niek understood what I felt and helped me to grow my possibilities as a diagnostician and creative therapist, enhancing my sensitivities and therapeutic possibilities. I adapted Niek's model, his methods, and his understanding of the nature of human beings to animals with great success. I began to feel and understand what an animal felt, and I was able to translate that for the owner. Often the owner felt something too but lacked the ability to use this information to make a bridge to their animal. I adapted Niek's physical exercises to animals, which resulted in more stability, better health, and happiness for the animals and, as a consequence, for the owner.

I continued to practice a mix of traditional Western veterinary medicine and Niek Brouw's method until 2017, when I gave up traditional medicine. This has given me the time and space to assemble these subjects and, hopefully, touch pet owners and animal lovers in a new way.

This book is a compilation of workshop summaries that give a brief view into the many ways that we can look at domesticated animals – what the possibilities for relationship with them can be, and what can go wrong. Each chapter is the distillation of an entire workshop, some four hours in duration, others 16 hours over two days. The people in attendance primarily wanted to live more harmoniously with their pets and deepen their understanding of their relationship with their pets (and themselves). My purpose in translating this work into writing is to give a window into the many ways we can look at – and experience – our animals. From there, we start a process of looking at our relationships with our animals in a new way: how we are, how they are, and how we are together.

Relationship Between People and Animals in Freedom and in Health

For thousands of years, human beings and animals have lived together for many different purposes. As we evolve in society and consciousness, we share limited common resources – space, lifetime, and our natural energy. The meaning of having an animal is different for different people on a personal, societal, and cultural level. In Western cultures, people increasingly have animals as a part of life satisfaction and relationship.

When we live with animals by choice, we do so to achieve some sort of satisfaction, whether it is conscious or subconscious. Sometimes, we make the choice out of our feelings – it feels good to have a big soulful dog lying with us when we relax or work around the house. Sometimes, the choice is out of our spirit—to share movement and adventure. And sometimes, the choice is to satisfy our form – we feel more complete when we have an animal that looks a certain way or shows the capacity for acquiring and expressing knowledge. (They do well in the show ring or are competent to learn what we want them to learn. Dressage horses are a good example of this.) When we have relationship on all of these levels, when we can mix all of our different parts in relationship with our animals, we grow, and they can grow. We enhance our lives, and our experience of being human animals on earth.

As society has evolved, we have become more involved with knowledge and the development of technology. *We are often consumed by our responsibilities in a way that isolates us from our own nature. Our domesticated animals are susceptible to our influences but tend to be more connected to their natural selves than us human beings*. They adapt because they believe or feel that their survival depends on it, or in other words, they feel like they have to. Sometimes, their survival does depend on (ultimately) unhealthy adaptation. We are often more powerful than they are – it is our world. We can suppress them with limitations on all levels. Our relationship might be limited to movement, exercise, and play (the spirit) or focused on training and manners, rewards for learned behavior, and performance (mind/form). Or we may only have an emotional relationship, one based on the exchange of feelings (the soul body). All over the world, cats, dogs, and horses provide us with companionship, and the differences between the species, breeds, and individuals, and their relationships with us can give us insight into all.

To explore our evolving relationship with animals, it is helpful to recognize basic characteristics that are natural for them. All animals share possibilities, but not equally. Dogs can learn the behaviors that people ask them to learn – or force them to learn. The same can also be true with horses. Cats are unique. They love people but, for the most part, will not suppress their nature for their human companions.

Cats are fundamentally more connected to their emotions and less likely to adapt for people by suppressing their feelings or movement. Dogs are more connected to knowledge and rules; they learn behavior for specific types of relationship with people, most often based on knowledge and rules and often facilitated by reward. Often they suppress their natural feelings in order to satisfy the desires of their people – to do what they think their owners want them to do. Both dogs and cats enjoy natural movement, and it can be easy for that to be suppressed by our ability to control their environment and, if possible, their behavior. Horses live in our world for our purpose, often isolated emotionally and suppressed in terms of free choice for natural movement.

A natural, healthy relationship requires healthy, natural development. This means that we – human animals and all the other animal species -- have our real needs met when we need them to be met. Safety is paramount in terms of protection from the outside world, provision of food as truly needed, and emotional connection.

In Western society, developmental lack of safety is rampant in animals. They are often abandoned, separated from their mothers too early, without adequate nutrition when they are hungry, and subjected to life-threatening elements. Those experiences can leave an animal insecure and unstable. This can – and usually does – result in difficulties in their relationship with us.

Because we have our own developmental disturbances, it is possible that we can meet our animals in a way that allows us to both heal and grow, or we can experience a painful and destructive collision that costs energy and disturbs the relationship.

As we understand our animals more deeply, we have the opportunity to grow and discern their true needs in the here and now, as well as the pain of their pasts. In the process, we can encounter opportunities to understand ourselves more deeply. The result is a beautiful growing of ourselves and our animals. When this happens, we accumulate energy – the feeling of being full, happy, and satisfied – rather than feeling depleted by the experience of the relationship.

The purpose of this short book is to give a feeling of what we really can meet with our animals and how we can recognize their true nature, and ours, more deeply.

Spirit-Power: Spontaneity and movement, the natural force in an animal.

Form-Surface: Appearance; boundaries, rules, and knowledge; the container of the being.

Emotion-Feelings: To feel and to express feelings.

Emotional Relationship with Our Animals

Many animal behaviorists and trainers recognize that animals do not care what we think; they care about how we feel. That is – how we feel to them. If we are angry, they can become insecure. If we are relaxed, they can feel safer to share love with us. If we are afraid, they can become afraid. The combinations are many.

Our domesticated animals are absolutely dependent on us for their healthy survival in our world. They cannot grow in their wholeness if they are on the street. We, as a culture, have just begun to experience the possibilities for relationship with our animals and what they can offer us in terms of discovering parts of ourselves that are unique to those relationships and required for those relationships to thrive.

We recognize our pets' incredible value to us. The evolution of how we care for them medically, what we feed them, and how we care about the quality of their days is tremendous. In Northern California, where I practiced veterinary medicine for 35 years, we have several MRIs and CT scanners devoted solely to pets. Many veterinary hospitals offer around-the-clock care, and specialty practices offer advanced diagnostics and treatment for a wide variety of illnesses. There is an abundance of daycare for dogs, dog walkers, and pet sitters. Pet stores carry a wide variety of organic and human-grade pet foods. More and more, people take extraordinary measures to prolong their pet's life and provide a quality pain-free existence.

But are we growing together for life satisfaction? Are we growing in ourselves and supporting our pets to develop their many potentials? The momentum of developments in pet care can actually isolate us from the nature of our animals—and from our own nature. *We use technology to treat them as a "machine" rather than in the wholeness of what they are –*

feeling individuals whose experiences in life impact their health.

As we administer to our animals and organize their lives, are we really giving them the right food for their development, for their emotional satisfaction? What do we give up by spending so much energy and resources on perfecting the experience of our pets?

In my experience, animals and their people want and need emotional relationship but often don't know how to get it. They might feel something is missing but can't give it a name. There can be low-grade stress. Pets can be emotionally isolated. We can feel this isolation without really knowing what it is, and so we cannot effectively work with it. Sometimes, there are overt issues, such as destructive behaviors, that we really don't understand. We tend to try to look at current circumstances and make adjustments or adaptations to "help" the animal or create a solution. These types of acquired systems sometimes work but at the cost of an ongoing output of energy. Sometimes, these solutions don't work at all. We adapt to accept the behavior of an animal that we find difficult to live with. Ultimately, we can feel resentful or less connected to the animal, and stagnation can develop in the relationship.

In my work with animals, I have always been aware that I feel comfortable and satisfied just to be with them. I don't really need them to do anything; I am happiest when they are happy and satisfied. Because of my makeup as a human being, I feel a lot in the animals I encounter but have not always been able to translate my feelings. I feel understood by many of the animals that I meet, recognized for my emotions. And I have seen this in many of my clients' relationships with their animals. In our culture, we tend to focus on what we can give words to, understandably. But so much of our life experiences, especially in relationship to our animals, are non-verbal.

There are many meaningful consciousnesses for which there are truly no adequate words—gratitude, humility, equality, belonging, and more. By surrendering our thinking (what we already know) and surrendering to what we feel (allowing ourselves to have a pure experience), we can experience more in our animals. As a result, our connection with them is more complete.

Early in my career as a veterinarian, I was fortunate to begin working with Niek Brouw of the Netherlands and study an approach to working with animals that enabled me to get underneath the manifestation of their problems. Dr. Brouw developed a body of work that addresses the development of motor skills that begin in the womb and continue post-natal, and the consequences of developmental gaps, the circumstance where the animal is unsafe or not getting their basic needs met, that become the foundation or nucleus of an illness. Applying psychomotoric exercises to animals, I began to see an increase in their stability and health and disappearance of their emotional, psychological, and physical issues

Psychomotoric exercises, developed by Niek Brouw, restore the optimal tension in an individual, the optimal tension *for* that individual. The exercises work with muscles, bones, and joints. The result is restored function, happiness, and overall emotional well-being.

It can be a good exercise for us to watch:

What emotional states do we accept in our pets?
Are we dependent on them being happy?
Are we disturbed when they are sad or angry?
Do we try to change how they feel?

Most of the time, when animals experience an emotion, it is necessary for them to do it in that moment. To interfere with their process can suppress the animal, and they can become frustrated. In fact, they can become ill. What is optimal support for them in these moments? By staying anchored in ourselves, remembering that we are safe, that circumstances are safe, and allowing our posture and attitude to reflect this, we can meet the real needs of an animal in their emotional experience. In other words, when an animal encounters fear in a situation that is safe (for example, those animals that get nervous riding in the car), the radiation of our relaxation and trust supports them to relax and feel safe under the circumstance.

The Being Is the Purest Part of an Animal

Animals have their own unique nature, not just as a species or breed but as individuals. In nature, basic fundamental characteristics can be identified on the levels of mind, soul, and spirit. Is an animal naturally extroverted or introverted? Are they naturally more emotionally expressive (physical displays of happiness, angry vocalization, or expression of sadness), or is it more natural for them to quietly and peacefully feel where they are and all that is in their surroundings? Are they creative by nature? Are they passive by nature?

Behavior, when it is pure, is a natural outward manifestation of the being. When development is healthy, behavior is healthy and is a pure reflection of the individual's being. When there is a disturbance in the natural process of development, behavior can—and usually does—result in an acquired system, an adaptation to the disturbance, an adaptation to survive. These behaviors give an advantage relative to the circumstances that existed at the time but ultimately become obsolete, unhealthy for the animal, and possibly destructive to the outside world. When there is an understanding of these adaptations, we can recognize and heal them in the here and now.

A very high percentage of animals have severe development disturbances. One of the most common is abandonment shock. Abandonment is common in young animals. Shelters and rescue organizations are full of young animals that are found stray or that have been surrendered. Animals —like humans—require safety for the development of inner security, stability and ultimately for healthy expression. It is important for dogs and cats to remain unaware that there is such a thing as unsafety until they are at least four months of age. If they become aware that unsafety exists and that they are unsafe, natural development is interrupted, and survival mechanisms come into play. The behavior becomes a reflection of "what do I need to do to survive?" Certain destructive behaviors—aggressive behavior, separation anxiety, and elimination problems can be understood from this point of view.

As animal owners and lovers, we have unique relationships and experiences with our animals. Are these relationships healthy, and do we really know who are pets are? Is their behavior a healthy natural manifestation of what they are, or is it a camouflage for their true nature?

Animals Are Fundamentally Emotional and Feel Life

In the Western world, we define health by analyzing fitness, physiologic and biochemical parameters, and behavior. In the natural world, emotional freedom, life satisfaction, and purpose are hallmarks of health...they are characteristics of a healthy animal.

Natural healthy development requires specific conditions to exist. When natural conditions are not met, for example, when there is a lack of safety for a young animal in the nest, that animal can feel threatened, and a development disturbance occurs. Normal development that would naturally occur at that moment stops, and a bad memory is created. The animal then lives in relationship with the world and itself, influenced by that bad memory and without the foundation that would have otherwise developed. The effect of development disturbances has not been integrated into the practice of medicine on a daily basis. Yet subtle or overt development disturbances can create chronic tension and anxiety that delay or prevent the resolution of medical or behavioral issues. Development disturbances range from overt physical trauma to subtle emotional or psychological trauma.

In our daily life, we have many responsibilities and tasks. Our society does not place an emphasis on feeling or creating a foundation of safety for happiness. This has far-reaching consequences. In the care of animals, there is an inherent purpose for happiness and emotional satisfaction for animals to be their nature. Feelings of satisfaction, contentment, and peace are natural aspects of emotional, psychological, and physical health. ***In a healthy animal movement and free choice are based on feeling. Happiness is, in a pure and simple way, the nature of animals.*** Our approach to caring for animals is task-oriented. We *do* a procedure or *prescribe* a course of medication to alleviate a situation. It's a "problem-solution" orientation. This seems reasonable, but there are other considerations.

A psychomotoric model or approach looks at patients and feels where they are strong and weak. An animal may present to a veterinarian for vomiting. Is the animal relaxed or tense? Does the animal feel safe or unsafe? Are these situational or chronic? Is the animal owner providing an environment conducive to healing?

In the world of animal care and animal relations, we have created sophisticated methods for analyzing and treating disease and behavioral problems. These modalities are primarily mechanistic and often based on control. They are useful and extremely important but rarely touch the animal on the level of the development disturbance in a way that allows them to heal and grow a stronger foundation.

Mind, soul, and spirit have been identified as fundamental characteristics of human beings. The same is true of animals.

These parts develop at different times and ultimately integrate and work together for healthy functioning, which includes happiness and emotional satisfaction. Every animal is unique and can experience a particular development disturbance in a different way. ***When an animal has had a disturbance on one of these levels (mind, soul, or spirit), they can heal in the here and now.***

Shock, Hunger, and Envy

Shock, hunger, and envy in animals are mechanisms that are the result of early trauma and create disbalances, behavioral problems, and, oftentimes, illness. Some can survive in our society without having the commitment of a human being, but their existence is certainly not optimal for growing, happiness, and thriving. When they are not part of a human family structure, most animals die at an untimely age. They depend on us to survive – and thrive.

Love for animals and commitment to them is strong in our culture. Our pets bring nature, love, and a connection that helps us feel ourselves, what we are, and what real, natural relationship with life on earth is outside of society's norms. (Although our pets are falling prey to the projections of our societal norms.) But as much as we cherish our animals, they can bring us trouble. We can be uncertain about how to best care for our pets; we can suffer from their behaviors that include urinating inside our homes and aggression towards other animals and people. Animals also exhibit impulse-compulsion behaviors that distress and upset us, disturbing our possibility to relax in ourselves. Some of these behaviors lead to conflict with friends and neighbors.

Animals are innocent. Our pets are born on earth dependent and pure. They need safety for healthy development, just like us. In a healthy individual, there is harmony and happiness. When our pets don't get their needs met, they cannot grow in a healthy way, just like us. For example, when kittens have been neglected or abandoned, they hide or isolate themselves; they have mistrust and will only come to us if we give them the space and time to do so. Our pets can develop many other disturbing and/or destructive behaviors that can make harmonious relationship impossible and lead to their surrender or demise. Sometimes, they can develop overt illness.

Abandonment, in particular, creates a circumstance in an animal that results in jealousy, judgment, feelings of being a victim, and revenge. They feel in another what they have missed during their development, and a feeling to harm or destroy the other can result. This sensation is a reaction. Strictly speaking, there is no real emotion in this. This mechanism produces an unconscious behavior; there is no here-and-now reality in it. It is a destructive reflection of the animal's past.

Who Are We to Our Pets, and Who Are Our Pets?

To be ourselves, to be our nature, is something we experience with our pets, sometimes better or more easily than we can in society. Our pets understand nature when they feel it – a cat lying on its back basking in the sun, a horse running through a field, or a Labrador swimming in a lake. They also understand or identify the nature in us when they feel it. Here, we can connect with them and develop a healthy relationship together. Animals allow us to be more ourselves than society does; they do not ask us to adapt or conform. In fact, they love it when we are connected to our true nature.

Society has trained us to conform and think in ways that are often based on fear and exclusion of some of our natural functions, including healthy instinct. Animals recognize this, most often subconsciously—they can be confused by being forced to go in a direction that is oppositional to their instincts.

Sometimes, we feel a lack of connection with our pets. We might feel that there is something wrong. Our pets might exhibit behaviors that are destructive to our environment or painful to us because the behaviors make visible their unhappiness. So often, we don't know what to do; we don't know how to reach them or reach that part of ourselves that can find a solution.

Our societal methods, behavioral and medical therapies, are based on control, not on finding a solution where the problem is. We might be blocked, or they might be blocked from engaging with each other in natural ways. Some people will avoid an approaching dog out of fear when that dog is kind. Some people will approach a fearful dog that can become aggressive, believing that they can help that dog. It is not their fault. It's not our fault. The problem is rooted in the past when requirements for natural healthy development were not met. There are times when our inability to solve a deep problem in

ON LIVING WITH PETS

our pets is simply a lack of experience and the natural consequence of living and growing together. The experience of discovery involves finding what is underneath or at the root of a problem and how to work with it. Being together, without judgment or feeling like a victim, is key.

Mind, Soul, and Spirit

Mind, soul, and spirit have been defined in many different ways. I use the definitions given by Dr. Brouw and the anatomical correlates. Different animals, by species, breeds, and individuals, have unique makeups of mind, soul, and spirit.

- The mind part is the form of an animal, their surface, their ability to accumulate and express knowledge, boundaries, and rules. It is a container for knowledge that defines the form of the animal. In the Dutch language, the mind part is called the body, as opposed to the soul and spirit. Our society in the United States largely focuses on the mind part.

- The soul body is the emotional part of an animal, their ability to feel and express feelings.

- The spirit is the spontaneous part of an animal, their power and movement.

When an animal is a feeling animal and the owner wants a companion for movement, the feeling part can become neglected or suppressed. For example, Great Danes are emotional introverts If forced to be social as an extrovert or forced to perform in a racing situation, they may become angry, depleted, or ill. A highly spirited animal like a vizsla or a whippet may become depressed or develop skin or neurologic problems if not given the freedom to express their spirit, that is, to run and play in freedom. There are many possible examples, the result being that an animal may become ill and isolated in its feelings. They may become depressed or develop neurological or gastrointestinal issues .

Psychomotoric exercises developed by Niek Brouw can facilitate the recovery of the natural state of the animal and result in the resolution of a disease process.

Our pets provide us with relationship opportunities and enhance the quality of our lives. They also bring us manifestations of their development disturbances in the form of difficult behaviors and/or illness. Our pets are as sensitive as we are and they are affected by their environment, including the subtle (and sometimes not so subtle) energies of the world we live in.

When natural development is disturbed, the seeds for unhappiness, lack of peace, and disease are sown. Animals (and people) adapt when they've had a shock and use their healthy parts to compensate and cope, but ultimately, these adaptations fail, and the animal can experience a crisis or disease. Circumstances, including external energies, can trigger or expose the shock. and as a result, behavior changes or illnesses appear. This scenario is extremely common and is under-emphasized in our Western models of treating disease. Yet, recognizing it is essential for healing and establishing a good foundation of health and happiness.

The here and now is the consciousness that can allow an animal to feel, in a relative way, that they are truly safe. Then, they can experience freedom in themselves and resume a healthy, natural growing process. Before here-and-now consciousness manifests for an animal, there can be an awareness of vulnerability, an unknown fear can exist that drives behaviors that are unhealthy, disturbing, and destructive. Through an understanding of natural development and what is necessary for safety in the self and in the outer world, difficult problems can be solved. *Emotional safety is under-emphasized in our Western models of treating disease.* It is essential for a good foundation of health and happiness. When we are in our trust and love for our animals, we show that we believe in their possibilities and capacities for change. In these moments, our animals touch the experience of freedom in themselves and resume a healthy natural growing process.

We – animals and people – are living beings who grow. And to grow is to transform on all levels: our feeling part, our spontaneous part, and our mind part. When we can't grow, or we don't grow, we can become ill. Working with animals and having a natural relationship with them that is born out of love and freedom, we can grow together, create, accumulate energy, and have more solid, lasting health and happiness.

To be in the here and now optimizes our ability to feel what is happening with our pet. To feel is fundamental and gives the most accurate information about a situation. One way for us to enhance our here-and-now consciousness is to sit on our hands, with our sit bones in the palms of our hands, relaxing our trunk, letting go of thinking and staying in that position for a few minutes.

Balance and Stability: Working with the Muscle Chains in Dogs

Bringing a dog into the family has purpose and carries with it an expectation to enhance our lives with relationship – companionship, love, and play. There is an exchange of energy that can be hard to describe.

Every individual dog has their own unique balance, their own unique optimal tension in the body, and their own unique nature—even when they're all German Shepherds, Labradors, or Chihuahuas. When dogs are alienated from their nature, out of balance, their possibilities for experiencing peace are diminished. They can experience panic, fear, anger, and sadness. They engage in survival behaviors, doing what they think they need to in order to feel safe. As discussed in previous chapters, disbalances originate when the dog's natural healthy development is disturbed, and they can also occur when animals experience a shock (trauma) as adults.

Muscular function has a huge influence during development as well as in day-to-day life. *Muscles work in the body to produce posture and movement. They function in chains, in relationship to the individual's inner world and their circumstances in the outer world.* During development, unsafe circumstances can produce fear tension in the muscle chains, which can have an effect on tissue development (for example, bone shape) and result in behaviors that compensate for the gap in the development (for example, aggression and neuroses). Overt orthopedic issues and other illnesses can also occur. Working with the muscle chains, we can give an individual dog the experience of being in balance, free to be as they are. As they surrender to the new experience of balance, then over time, physical issues and difficult behaviors, including problems with connection in relationship to their people, can resolve. This can take time and repetition because of the tenacity of cell memory.

The Muscle Chains

In nature, animals develop in a healthy way; they are able to reach their potentials when their true needs are met. Sometimes, characteristic functions that would be naturally predominant in an animal are suppressed by circumstances. For example, when an animal that is an emotional extrovert is not allowed to express their feelings. Greyhounds are emotional extroverts and express themselves through movement. If their movement is suppressed they can become depressed. In general, suppression of natural function can result in an unhealthy tension in a muscle chain, restricting natural movement and putting stress on the bones, joints, and connective tissue. This can make an animal psychologically and/or emotionally unfree. A dog that is a social extrovert who is forced to behave as an introvert, can over time develop orthopedic, emotional, and/or psychological problems because there becomes a fight between the antagonistic muscle chains, and rigidity develops in the system.

It is also possible that a young animal experiences a trauma, such as abandonment or abuse. As a survival mechanism they can come to rely on a particular muscle chain. For example, too much social introversion might be safer for their circumstances. That can dominate the quality of their movements, resulting in overuse of that muscle chain and underdevelopment as a result of underuse of others.

Traditionally, muscular function has been characterized by voluntary or skeletal muscles, involuntary (autonomic), or smooth muscles with some other variations. Skeletal muscles are the subject here, those muscles that we use to bend, stretch, move, etc. In veterinary medical education, skeletal muscles are studied in terms of where they originate, where they insert, and what specific function they perform. (For example, the biceps muscles flex the elbow.) Natural movement is not an isolated function of any particular muscle. Movement is the result of many muscles working together, in relationship to each other as well as other parts of the body. When we are healthy, our movements are in harmony with our feelings and intentions, our thoughts and our desire to reach a target. Usually, our movements are influenced by what we feel, what we know or don't know, and where we want to go.

Complex movements, initiated by deliberate conscious choice as well as movements born from the subconscious, are executed by muscles working together. Work done in Europe by Godelieve Denys-Struif and Niek Brouw demonstrates that muscles work together in chains to produce different postures and different types of movement. Social introverts have a posture that features internal or endorotation of the limbs. Social extroverts have a posture that features external rotation of the limbs. Emotional extroverts tend to have more kyphosis or upward arching of the column, and emotional introverts tend to have a slightly forward bending posture. The work has been published, to my knowledge, in French, Spanish and Flemish. I have not come across an English-language version.

As a young veterinarian I became interested in why animals developed certain disorders. I wasn't satisfied with the mechanical and geometrical explanations. Why did some animals tear a major supportive ligament in the knee and others under the same circumstances didn't? And what's underneath the why? What led an individual to develop a conformation and manner of moving that caused the premature deterioration of structures in the body? The answers to these questions are complex and require the recognition of the six basic muscle chains in the body and their relationship to posture, movement, and psychological and emotional expression.

I started to work with the muscle chains using the exercises developed by Dr. Brouw in 1989. My practice experience has been almost exclusively with dogs and cats, with equines becoming an increasing part of my work. These species have, basically, the same muscle chains as human beings. And the muscle chains in animals have the same essential functions as they do in people.

Typology

The six muscle chains exist as three pairs of antagonistic chains. They allow an individual to bend and stretch, internally and externally rotate the limbs, and move in a horizontal or vertical trajectory. Different breeds have predilections for particular muscle chains, having a dominant development in the individuals within their breeds. For example, golden retrievers are natural social extroverts; sighthounds, as a general rule, are social introverts. Great Danes are emotional introverts, and greyhounds are emotional extroverts. Draft horses have a predilection for horizontal movement and Arabian horses have a natural predilection for movement that has a component of vertical trajectory.

There is a relationship between skull shape and the natural predominant muscle chain in an individual. Dogs, as a species, have an enormous variety of breeds with huge differences in skull shape, as well as body shape and size. The result is a variety of posture and movement possibilities and a wide variety of breed-specific personality traits.

Corrective Exercises

When circumstances have resulted in the unhealthy development of the muscle chains, the tension in those chains comes into disbalance and can create unhealthy posture and movement. Exercises have been developed that can bring healthy natural tension and freedom into a muscle chain. When this happens, the animal will most often have an experience of freedom. The experience may feel strange to them, unknown, and sometimes they can become disoriented or confused. These are conscious states that are a natural part of coming out of suppression and into balance.

The Six Muscle Chains

Most simply put, there are three pairs of muscle chains. Each muscle chain in a pair has an opposing function.

- Social function: social introversion and social extroversion or expression

- Emotional function: emotional introversion and emotional extroversion or expression

- Spontaneity, movement, and power (the spirit): horizontal movement and power, and vertical movement and power

Dogs, cats, and horses have all of these muscle chains with some variation in the components. Species, breed, and individual differences contribute to the individual's personality and function possibilities.

Dependence and Independence

Obviously our pets are dependent on us – they cannot survive in our world, in our society, without us. They need protection from our cars, from natural predators (some of our pets have lost their natural instincts or have them suppressed in their deep subconscious), from society's rules (for example, restriction from free play at the beach or in the forest), and from people who judge them for certain behaviors or breed characteristics. Animals also need us for food, shelter, medical care, and relationship. We have a responsibility to satisfy the true needs of our pets when we accept and commit to the relationship.

Pets are dependent on us for safety in the outer world. Some animals are dependent on us for safety in their inner worlds. We are often dependent on our animals for safety in our inner world. It is rare that we are dependent on our animals for safety in the outer world.

But what about unhealthy dependence? At what point do our own shock parts and vulnerabilities cause us to negatively influence our pet's experience of the here and now? How do our own unsatisfied needs – those places where we have met pain in the past – distort our perceptions? And how can this interfere with our pets' natural development?

When we satisfy or are completely responsible to our animals' dependencies, we can meet our own shock parts, touch our own unsatisfied needs, where we are still vulnerable. Sometimes, we can end up resenting our responsibilities or suppressing the natural behaviors and expressions of our animals. Or, as a projection of our own needs, we can over-give. As a result, we can end up in unhealthy relationships with our pets. We can end up living with difficult behaviors such as inappropriate urinating, excessive barking, or aggression.

When we live naturally in harmony with our pets, they do not need us to process their emotional experiences—they

only need the safety to do it. Commonly when an animal has an emotional expression or radiation that is fear or sadness, pet owners want to help them with that feeling. So often, helping is really hurting, it prevents the animal from processing that feeling completely. All they really need is our stability, the radiation of our love. Certain of our animals' feelings trigger our own shocks, our own unsatisfied hunger, and other unhealthy triggers such as paranoia or guilt.

Natural development requires that animals get their basic needs met in safety: food when they are hungry, protection from the outside world, love, and confirmation that they belong and are welcome to be in this world as they are.

Consider:

What is my pet's behavior bringing up in me? Embarrassment? Fear? Anger? Sadness? Paranoia? Guilt? Compulsion?

What happens in me when my pet's behavior does not match my expectations or needs? Or, in other words, when my pet's behavior disturbs me?

Perception and Relationship

Natural relationship between our pets and ourselves is a foundation for good health and growing for all of us. These relationships provide a structure to help us discriminate between healthy and unhealthy behaviors. Animals have their own unique make-ups and personalities that can be very different from societal norms and expectations.

We all have had shocks that disturb our development which can result in gaps in our consciousness and our perception possibilities. As human beings, we can be suppressed in our natural intelligence, instinct, intuition, and emotional intelligence. We can be suppressed by lack of caressing, lack of natural breastfeeding, restriction of natural movements, and by lack of the feeling of safety or lack of actual safety at any time.

We are responsible for identifying these gaps in ourselves and in our pets. How to recognize when we have a restriction of our perceptions is a conundrum. It can be easier for us to do that for our pets.

We see the reality of certain situations in ways that they cannot. For example, fireworks. Our pets might and most often react as if they are in imminent danger. But we know that we and they are not in danger. We are safe despite the noise. When we consciously identify their misperceptions, we can create new bridges of understanding to the here and now. We do this by reflecting through our posture and attitude that there is no problem; in essence, we don't react. We relax and take the noise in stride. We are then supporting our animals in their growing processes.

Development of our feeling part is an important asset – when we can feel what our animals need or why they engage in a particular behavior, the information is usually more accurate than analysis. For example, animals urinate in the house for many different reasons. If we feel that our pet is insecure and feels vulnerable then we can begin to explore and work with that problem.

Every animal is unique in their make-up as well as in the shocks that have suppressed them. So are we. Therefore, there's no formula or single set of guidelines for developing a natural, healthy relationship. But we can create a structure for understanding where we are dependent and where we are independent as influenced by our own shocks and disturbances, addressing the same questions for our animals. This is fundamental for creating healthy relationship, healing, and moving forward. Having this kind of understanding gives us all a foundation for solving problems that occur commonly with our pets, including aggressive behavior, elimination problems, and lack of emotional connection. We then can become more independent in how we work with our animals.

How Shock Manifests in Our Animals

Some animals are healthy. Some animals appear to be healthy or healthy enough to live in harmony with their owners. Some animals have health issues that are dormant and that, over time, become explosive. There are often tell-tale signs, feelings of isolation or paranoia. An unhealthy relationship with food, for example, often manifests as a power game around eating and can lead to obesity and diabetes in some animals. For animals that refuse to eat or are extraordinarily selective, it can also result in chronic poor doing (weight loss, low energy, poor hair coat) as a result of insufficient nutrition.

Many of our pets have disbalances in their natural selves that end up in either overuse or underuse of a natural system in the body. When an animal is not allowed to exhibit a natural behavior or denied the freedom to feel what they naturally feel, then that particular function becomes suppressed, and an unnatural tension or unhealthy pressure can build up in the body. This unnatural pressure caused by suppression of emotional expression and lack of natural boundaries can result in tissue destruction – actual hypertension in the circulatory system, gastrointestinal distress, or ulceration. Hypertension, or high blood pressure, is common in cats without any external evidence of stress or disturbance until their bodies can no longer contain it or compensate for it. These cats often do not have freedom *to be* their emotions or to express their emotions. Cats need to give love and have it received, and they need to be given love and receive it.

Many of our pets suppress their natural movement because of conscious or subconscious fear caused by a here-and-now problem or a shock/ disturbance from the past. If chronic, these suppressions can result in joint disease or psychological depression. I have seen animals present for apparent illness – not eating, lethargy – and the underlying cause was a lack of function from suppression of a particular muscle chain in the body or lack of normal rhythm in the abdominal organs.

Many problems presented as health issues are the result of development disturbances or lack of freedom in an animal's day-to-day life. These disturbances produce symptoms that lead pet owners to seek care for their animals. Very often, the symptoms are investigated/addressed, defined in terms of scientific models, and then treated or managed with medications (a type of control) or other mechanical solutions. In some systems, for example, homeopathy and Chinese medicine, energy disbalances or blocks are identified and treated. These solutions can be successful, but sometimes they are temporary or fail altogether.

When there has been a development disturbance, a natural system (or systems) in the body is not operating at its optimum. Many drugs are used to control the resulting symptoms. Often, their success is temporary, and the drug's ability to manage a patient to its natural life expectancy fails. Animals often resent the administration of medications because they don't understand what's going on since there is no bridge or understanding or because the act of administration triggers a shock. Optimal quality of life, real growing, is not restored. Heart conditions, for example, are commonly managed with medications, and as time goes on, more medications become necessary until, in many cases, the patient "unravels," succumbing to their disease. Animals can appear to heal with medications (sometimes they do), but often, a new disbalance will emerge.

Illness presents an opportunity to improve health and tune an animal for natural growing in the future.

Equality and Relativity: Feeling Our Pets and Ourselves and Understanding Each Other

All animals have a language for exchange – vocalization, body posture, radiation, movement, touch, and feeling. These are some very basic forms of communication. In the world of animal and human relations, there is an abundance of communication on many levels. In my experience, we are engaged in it all the time. Sometimes, when we are away at work or on a trip, we think of our pets. They can feel that. Sometimes, we start thinking of them because they are thinking of us or feeling about us. Non-verbal communication between human beings and animals is huge and can travel distance.

We are all equal as living beings on earth. Because of our developmental differences and a lack of anchored consciousness in our equality, we can lack the right foundation to enjoy the feelings of being together. *We have different capacities for survival, and we need each other.* Looking at who we are as animals and human beings, we can begin to experience each other in a relative way, in our emotions.

Our pets need us to survive in our world, and this does not make them unequal to us, they are different from us. As a result, we have natural responsibility for them. Some people are more dependent on others in a natural way because of handicaps on different levels of natural development. The same is true with animals – they are each unique and have different dependencies on each other and us.

Our perception of animals in relation to us is healthiest when we consider their behavior, posture, and attitude in a relative way, that is, to consider all of the circumstances that may be in play at any given moment, including their past. To be in the here and now, to feel our natural equality of existence, can be a foundation for growing our understanding, love, and satisfaction in living together. Judgment of our level of development, of our animal's development, or our animal's behavior only interferes with our capacity for love and understanding.

Being together without judgment opens the possibility of experiencing freedom and peace. Animals understand this very well and often are more connected to these feelings than we are. They seem to have more time to feel, ruminate, and digest their experiences. *We live in each other's energy, and we have an effect on each other. To become conscious of each other, with respect, is a wonderful way to grow.*

Touching Animals to Change Their Consciousness

When we touch our pets, we affect them. Sometimes, when we touch our cats they start to purr, and sometimes when we touch our dogs they wag their tails or lick us. Touching is a way to exchange affection. There are many ways we can touch that have far-reaching effects that can change the way an animal experiences life.

The skin is a large sensory organ that receives information from the outside world. The deeper tissues, the connective tissue, the bones, and the joints also receive information and, as a result, give information that can change the way an animal feels and behaves.

Often, when people touch their pets, it is a reflexive gesture born out of love for them, a type of automatic system. They enjoy it, and, at least in a subconscious way, it makes them feel safe. We can also touch more specifically and consciously. We can contact their bones when we gently touch with our bones; this can result in more freedom of movement and more choice in movement. The same is true when we touch with connective tissue (the tissue between the skin and the muscles, using smooth, gliding touch), muscle (there are a myriad of ways to use our muscles to touch theirs), and joint consciousness (a light dancing touch). The result can be that animals start to move from their own feeling. They are then free to move out of choice and not only in reaction to circumstances or triggers in the outer world or a difficult or overwhelming feeling or sensation such as fear or panic. Animals then begin to originate movement out of their true natural selves.

What is our motivation to touch our animals?
Do we bring the energy of wanting to help them?
Do we do it to soothe ourselves?
Or is it to share love, to exchange feelings?

To grow in our awareness of our motivation is a process. We begin by becoming aware of our intention as to why we want to touch our pet, feeling how it is received and how we are after we make our choice.

Accessing the Unknown in Pets

Natural growing occurs in all of us and our pets. Living with animals is a choice we make because we want to enhance our lives, our experience of being alive. In a turbulent and difficult world, we can easily lose contact with our nature and purpose. Animals can remind us of what we are in a feeling way, and help us access ourselves in a way that is often not available in our mind-oriented society.

Our animals are fundamentally as we are. Most of our pets have been shocked. Rescue animals are, by definition, abandoned. Deeper to the abandonment, in most cases, are other shocks – unsafety, of the mother, during pregnancy as well as in the nest, lack of food, illness, or death of a sibling. These animals develop strategies and behaviors to survive that are not a reflection of their natural selves. Although these survival mechanisms may be effective for a period of time, ultimately, they become obsolete and can create illness or behavioral problems later in life. Sometimes they can't find an adequate coping mechanism and engage in irrelevant or destructive behaviors.

What is the true nature of your pet?

Most of our pets exhibit behaviors, radiate energy and display postures that are a reflection of development disturbances as well as their true nature. When there is a shock, there is an underdeveloped part, full of potential and beauty, and it is unknown to the animal and to us. Accessing their true nature is accessing the unknown.

The unknown is just that – we have belief systems about what relationship with a dog should be or look or feel like, and often, those beliefs represent only a small part of what is possible. Working with ourselves and our animals to heal shocks and access the unknown is natural growing. Discovery and healing of our hidden parts and unrealized potentials are happiness and health.

What Your Animals Need and What Your Animals Want. What Do You Give?

When we are faced with difficult behaviors and strive to meet all of our animal's needs in the here and now, often our pets are not satisfied or happy. They engage in behaviors driven by needs that we feel should be satisfied by what we give.

When we adopt animals, they have had a transition from a healthy or unhealthy nest period to us. Most of the time, the nest experience is sub-optimal. Most young shelter animals have not had a safe nest experience; they have been abandoned and likely traumatized from lack of food and safety. As a result, certain types of behavior can manifest—apathy, rebellion, and revenge, to name a few. It is common to expect that if we satisfy the here-and-now needs of an animal they will behave satisfied, happy, and healthy. But they don't! Because of their unhealthy experiences at an early age during a sensitive developmental period, they are protective of their shock parts. So when they ask for food in a way that reflects their shock, not their here-and-now reality, it is unlikely a free expression. And we can become bound into an unhealthy dynamic. When they receive the food (food food, love), they often do not have a natural experience of satisfaction.

To recognize what an animal needs in the here and now for optimal health requires perception training and an understanding of normal healthy development. Then, it is possible to recognize what an animal really needs, and you can give it.

The Hunger Mechanism: Why Do Dogs Become Aggressive Around Food in Our Homes?

Resource guarding is a popular term that is used for dogs that display aggressive tendencies in relationship to food. Again, the animals that I work with are very well cared for and there is no lack of food.

In a healthy animal, the hunger mechanism creates the feeling of hunger when an animal needs food. When the food is not there, the animal will ask for it in freedom, a free expression of "I'm hungry, I need food." A healthy asking is not demanding, a dog may walk to its food bowl and then approach its owner.

When the hunger mechanism is disturbed, there are two possibilities: apathy and rebellion. If food doesn't come after many attempts, and when the animal starts to feel that their survival is threatened, they can become very passive—apathetic in order to conserve energy while they wait for someone to recognize their need and feed them. Another possibility is that the animal becomes rebellious and demands food. This can manifest as aggression.

Disorders of the hunger mechanism can be very challenging and difficult to heal. One very straightforward approach is to feed the animal before they are hungry. This can take some trial and error and will often defy daily routines. But when a dog with a disturbed hunger mechanism that has a manifestation of aggression is fed before they are hungry, they appear satisfied; they will become disoriented and have an experience that reflects their reality – food is there. It can be confusing for them because they have a deep-seated attitude that there is not enough food. Allowing for the disorientation and confusion by staying relaxed (and happy if possible), the dog can come into the here-and-now and have a new experience of safety, security, and satisfaction.

Dogs whose hunger mechanism shocks manifest in apathy rather than rebellion or aggression are, in effect, attempting to manipulate their owners into giving them food with a helpless, pathetic expression that elicits sympathy and extraordinary measures to get them to eat. It's an energy drain. This perpetuates the behavior and the shock isn't healed. It's an unhealthy system.

Denying these dogs contact until they start to move for themselves in the direction of food can be healing. Sometimes, they will become angry. Discovering the most effective way to provide food in these circumstances requires creativity and trial and error.

The challenge is to stay in the here-and-now reality and not indulge the shock manifestation. We do this by not responding in any way to their unhealthy expression. This can be very difficult for a wide variety of reasons, particularly because of our emotional vulnerabilities as well as our own unhealthy relationship with food. We meet ourselves in this process. It's difficult, doable, and rewarding.

Psychomotoric exercises are very helpful.

Perception Training and The Soul Body: Solving the "Unsolvable"

So many animals bring disturbing behaviors to our relationships with them. The behaviors may be overtly destructive or subtle. Overtly destructive behaviors include excessive barking, property destruction, aggression, and neuroses such as constant neediness. We might notice that our animal isolates itself or we sense that they are unhappy. We might feel that the animal is not satisfied, content, or peaceful. We might not feel satisfied with the relationship.

Many pet owners know or have an idea of what they want from their relationship with their pet and how they would like a situation to change. It is understandable. Unfortunately, this vision often is not practical or realistic because it doesn't take into consideration the animal's unique challenges or their true, natural capabilities. When animals bring destructive or disturbing behaviors, it is likely due to shock, hunger, or envy. Engaging the problematic behavior directly in an effort to control or extinguish it can work, but in my experience, many problems can't be dealt with in that way. Many pet owners end up living with a problem that chronically costs energy and ends up in an unsatisfying relationship.

We are responsible for recognizing and assuring the safety of our pets. Allowing them to experience benign noise or commotion (for example, sirens or a loud truck) in safety and reflecting back to them relaxation and lack of concern can give them the experience that the noise is not relevant to them and they are safe.

Going into the unknown to discover the healthy nature of an animal is a way to grow a satisfying and healthy relationship. This means trying new things, self-reflection and self-examination with respect to how we are in relationship with our pet. Here-and-now perception is critical in order to see and feel what is happening in the animal so that we can provide the circumstances that will resolve developmental gaps.

What is the Relationship Between Your Relationship with Your Animal and Your Animal's Health?

Our pets are so dependent on us that we have a huge influence on their development and well-being for their entire lives. While their development is most sensitive when they are young, it continues throughout their lifetime as it does for us. We have opportunities to share and exchange, as well as to be alone, in harmony. For me, the feeling of being with my animals in a pure way (without thinking) is wonderful.

In having relationships with our animals we are responsible for recognizing when they need food and what food is optimal for them, when they need medical care, when they are in a destructive behavior, and more, and more, and more. In deepening ourselves – our feeling world, our spiritual awareness, and our satisfaction with our animals – we open growing opportunities for sharing love and exchanging feelings.

So often, I hear in my practice, "My animal must be taking on my stuff." It is more common that animals are suppressed by their owner and can become depressed or neurotic because they cannot be their natural possibilities. Restrictions of movement, being overly interactive, worrying about their health, and pampering by overfeeding are some examples of how our pets are suppressed. They also need acknowledgment, confirmation of their being, compliments to their personality, spiritual play (times when we're engaged in movement with them), and emotional togetherness. Animals have a sense of their true nature, and they often need confirmation. To confirm an animal with a compliment born out of a natural feeling of love can create stability for them. We can recognize their beautiful nature, their love, and harmony with nature. They have a tremendous sensitivity for our posture, intonation, and energetic radiation. (Do we radiate peace, anger, sadness, etc.?) In the most basic of terms, **what animals need in relationship with us for optimal health is that we recognize their needs at any given moment.** Sometimes, they need space; sometimes, they need to play; sometimes, they need to feel hunger; sometimes, they need affection; sometimes, they need to feel that they belong with us. Suppression of the spirit can create disbalances in the spinal column and kidneys. Suppression of feelings can cause stomach problems and heart problems.

Healing is a complex but doable process that starts with safety. It requires us to let go of the external suppression that results from our expectations and efforts to control and instead see and meet the animal's needs in real-time (that is the here and now). The muscle chain exercises are important because they give the body the experiences and freedom necessary for the healing process to occur.

Another exercise that can bring us into the here and now is what we call "butt walking." Sit on the floor, and use a carpet for friction. Propel yourself forward on your butt, move backward, and side to side.

ROB ERTEMAN

Cats and Their People. Do Cats Need Relationship with People?

For those of us who live with cats, why do we choose to do so? What is the attraction? Do we need them? What do we get out of our relationships with them?

Cats have been revered by some cultures for thousands of years. They were considered special, of a higher intelligence, sensitive, and god-like. Today, cats are beloved and an important part of family life in many cultures. They enhance and enrich our lives in ways that we can feel but can be difficult to put into words. Cats are connected to themselves, particularly their emotional and spiritual selves.

My experience suggests that emotional exchange is probably the biggest reason that we choose to live together with cats. We enjoy their company along with the basics, providing food for them and taking care of their needs. I have seen the love that cats share with their people, as well as the love that we exchange with them. Cats have an expression of love that is very direct – they will ask for affection, they will be affectionate, and they will choose to be close to us at times. We can have similar expressions of love (asking for affection or giving affection) through touch and loving radiation, but sometimes we are disconnected from that part of ourselves, and our expression can manifest in doing too much for them.

We can obsess about their health and needs. Cats don't need a lot of toys or pampering. They need freedom to be who they are, and at the same time, they need relationship. Sometimes, their connection to themselves and their commitment to their freedom are interpreted as not needing or loving the relationship. Actually, the two exist together, and this is a living, natural relationship.

Cats have been domesticated for thousands of years. In the evolutionary process, we have become connected to each other emotionally. Cats are truly dependent on us in order to thrive in our world and have a complete life experience. Without us, they can often survive, but it is survival, not living. Cats are often accused of being aloof, overly independent, and disinterested in what we want for them and from them. They have "minds of their own." *It is my experience that cats are simply better connected to their nature. They are less apt to adapt their behavior or suppress their emotions because it might be what a person wants or it might make a certain situation "easier."* They are deeply connected to what they are, and when they are pressured to be something other than that, they become neurotic or ill. They need us to provide certain necessities for their survival (food, protection from the outer world, medical care, etc.). They also need us for relationship, emotional exchange, and love. They need us to be ourselves.

Self-Awareness and Our Relationships with Dogs: What We Meet in Ourselves When Our Dogs Meet Their Shocks

Our relationships with our dogs (and sometimes with dogs that are not our own) are unique opportunities for us to meet a pure place in ourselves where we feel love, acceptance, and a deeper part of ourselves than we often feel in our day-to-day lives. Because our dogs don't judge us, we can be more open and have emotional experiences that are safe, rewarding, and beautiful. The purity and simplicity that is available with our dogs result in a strong connection or bond and often a very satisfying relationship. However, sometimes, there are problems in these relationships that stem from behavioral issues ranging from destructive behavior to anxiety, depression, and isolation. As discussed, these problems are a result of disturbances or shocks, unsafety during specific developmental periods. It is extremely common.

When dogs have been unsafe in the womb, they often don't have the foundation in post-natal life to process their experiences in a relative way, to feel themselves in the here and now. Some manifestations of fetal shock include anxiety, jealousy, and paranoia. Another common developmental issue for dogs occurs in the immediate post-natal period – that is unsafety in the nest. This post-natal period of nursing and experiencing life relative to others (the mother, littermates, and human beings) is a period of accumulating feeling experiences. When this becomes shocked these dogs can show similar behaviors to fetal disturbances but tend to be more reachable for training and experiencing the here-and-now safety as it is.

When the dog's behavior touches a shock in us, we might isolate. In our society, we sometimes disconnect our emotional beings from the process of working with our dog's difficult behaviors that interfere with happy, peaceful, and loving relationship. We look for solutions in the form of manipulating the environment, training, adapting our own behavior, and often medication. Sometimes, a difficult manifestation of a shock can touch an underdeveloped part of ourselves, and, as a result, we meet a type of incompetence to handle the problem effectively. We can be reactive. In general, we are not aware of our developmental limitations that can interfere with our effectiveness with our dogs. We often feel that our love and devotion should be enough to heal our dog's difficult past. Even if we do have experience and knowledge concerning developmental shocks, our own shocks can make us blind to a solution. And sometimes, we are just too close to our dog to see things clearly.

Forcing Our Dogs To Do What We Think Is Best for Them (or Us)

In our society, the norm is to pursue help for our dog's unwanted behaviors in the form of training, conditioning, and/or medication. Sometimes, these approaches work, at least in terms of minimizing or eradicating unwanted behavior. Sometimes not. Even when we do get a result that appears to be a solution, sometimes it may not be healthy for the dog because difficult behaviors – anxiety, isolation, depression, aggression, and more – often manifest as a result of a development disturbance.

When we ask or force our dogs to do something in response to unwanted behavior, for example, sit, stay, or lay down, there is a possibility that we are also teaching them to do it out of fear or an unhealthy desire to please. If that is the case, then we are controlling behavior that stems from an underlying problem. We are suppressing a natural emotional expression in them. If we are successful, then the dogs will tend to suppress themselves and behave in response to what we want or need at a given moment. Many dogs end up behaving better in anticipation of what they think we want from them. The result is an unnatural relationship with us and the outside world and suppression of natural healthy streaming (their natural flow of energy) in themselves and in their behavior to the outer world. (Of course, it is necessary to force or control our dogs for their safety and for damage control.) Development disturbances can happen in the womb, in the nest, during their puberty, as well as at any time during their adult lives. Most of the disturbances or shocks that drive a dog's difficult behavior occur during the womb or nest development or are genetic. Nest development includes the period of nursing, and the entire nest period is approximately 4 to 6 months.

When we force our dogs to stop a behavior or perform a behavior that we think might be good for them, we might be missing an opportunity to give them some safety. Usually, the driving force for unwanted behavior is feeling unsafe, either consciously or unconsciously.

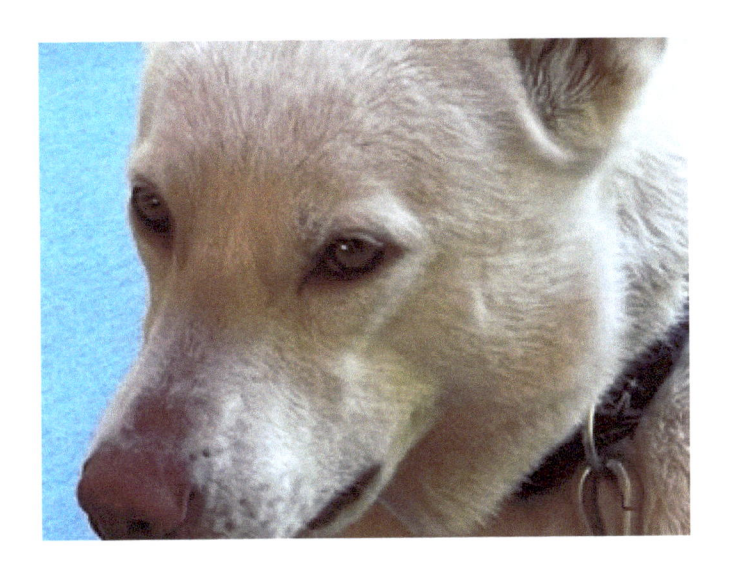

Healing Developmental Gaps as an Alternative to Training: Instead of Suppression, Freedom to Feel and Move

To exchange and share naturally with our dogs is healthy for them and for us. To be ourselves and for our dogs to be themselves is fundamental for healthy living together. The energy that dogs bring to us enhances our lives in many ways. There are opportunities for deep love, intimacy and experiencing nature together, just to mention a few. The known and unknown possibilities for relationship and experience are huge.

Because of developmental gaps in our dogs (and ourselves), the unknown can be inaccessible. It can be surprising how different a dog can be when they grow their unknown potentials. Until we explore more deeply why we might have issues with our dogs or why we are unsatisfied in our relationships with them, it is likely that we and/or they will repeat behaviors indefinitely.

When our dogs present us with challenging or destructive behaviors, our societal norm is to adapt to them or to train them so that we can control unwanted behavior. Sometimes, we adjust our routines or our homes to accommodate the insecurities that manifest in unwanted behavior. Sometimes, dogs will grow and heal as a result of our adaptations, but often, we end up in a routine that is unnatural and exhausting. Sometimes, our dogs end up suppressed in a way that is not necessary, and this can result in neuroses.

Giving dogs an understanding of what they feel (or what they can't feel) can result in a transformation. Understanding the seeds of a dog's unhealthy or destructive behavior can be the beginning of providing a feeling of safety that can result in them seeing and feeling things in a new way. Sometimes, this requires that we have a new experience in ourselves so that we change how we feel about them. The result can be that our dogs can feel the here and now as it really is. Most dogs that I have seen over the years are safe in the here and now, but they don't feel it. One of the important things that we can do for our dogs is to avoid pampering if they are afraid or panic in a safe situation. It is vital that we stay calm and relaxed. As they start to feel safe, the possibilities for deeper love and freedom, both the feeling of freedom and actual freedom in the outside world, start to grow. In the growing, the unrealized potentials or the unknown of the animal becomes visible.

Treating Aggressive Dogs Through Healing

Aggressive dog behavior is a common problem that is difficult to treat effectively. Aggression can be directed towards other dogs, people, and/or other animals, resulting in significant injury or death to their victims. Many aggressive dogs are euthanized. Statistics reported for 2023 vary tremendously and suggest that between 700,000 and 2.7 million cats and dogs are euthanized annually. (More information can be found on DogsBite.org.) These statistics only capture a fraction of incidents. In my experience as a veterinarian, I witnessed many dogs that were purportedly surrendered due to practical family circumstances, like a move. When I started to examine those dogs, I discovered that they were aggressive. It is not about blaming or judgment. Owning an aggressive dog is difficult, and many are put to death because of a lack of resources and/or know-how.

Treatment is normally based on training-control mechanisms such as avoidance and shock collars, conditioning, and counter-conditioning. Antidepressant medications are also used at times in conjunction with these methods. Spaying and neutering are used to minimize hormonal influences. These approaches, for the most part, address the symptoms of the problem but not the root. As a result, these dogs rarely become bulletproof, that is, they never become completely reliable. The nucleus of the negative behavior remains, and under the right (or wrong) circumstance, can be activated with devastating results.

Working with aggressive dogs using psychometric exercises can, *over time*, lead the individual to a resolution by healing the shocks that are the root of the behavior. The importance of the here and now has been discussed, and the administration of exercises to release chronic fear tension and to build up stability in the structures that anchor the here and now can actually result in the resolution of the behavior. This takes time.

Anxiety, Reactivity, and Aggression in Dogs: Finding Balance and Peace

Anxiety and reactivity are extremely common in our pet dogs. Aggression, or what we experience as aggression, is also prevalent but, fortunately, not usually dangerous. Nevertheless, anxiety and reactivity are destructive to the dog's environment, disturbing the peace in the family, often causing property destruction, and disrupting the environment inside and outside of the home. Often a dog's reactivity creates conflict between people and can result in ongoing stress for the family.

Anxiety can be defined in many ways – fear of danger or lack of safety when circumstances are safe, a feeling that something bad is going to happen, an unpleasant, nervous feeling in the body, unease, excessive apprehension of the unknown, repetitive apprehension of the known, and more. The seeds for anxiety often run deep. Causes can be genetic (also recently known as intergenerational trauma), intrauterine, and/or a result of disturbances when puppies are in the nest. These seeds exist in the subconscious of the animal as a deep "buried" memory that disturbs natural healthy development and essentially dominates behavior under certain circumstances. We call the event that seeds the unhealthy tension in the body and brings it into chronic disbalance a shock.

A shock is anything that disturbs the natural development or happy reality of the animal – illness, loss of the mother, neglect, abuse, etc. The shock results in a gap in the animal's foundation, leaving them unable to cope with certain benign and common circumstances in a healthy way. When the shock part is touched, it is, in effect, a trigger, and the animal reacts in fear, often defensively. This is a reflection of the past, not actual here-and-now circumstances. In my experience, most of the time, dogs are not reachable once a behavior is launched. In essence, they are reacting out of their subconscious, so they aren't aware of what they're doing until the surge that fuels the behavior begins to subside.

The solution is to build a bridge between the subconscious of the dog and the present moment to give a strong reflection of the here-and-now circumstance when the dogs are receptive to receiving or feeling it. This approach provides tools to bring dogs into here-and-now awareness that can relieve the chronic tension in the body that keeps them in disbalance. When in disbalance, they are unstable and unable to weather a storm effectively. So they become afraid and react. There are physical exercises that facilitate an individual to come out of the suppression from the past and into balance, resulting in stability so that they become relational in a healthy way. There are physical exercises to bring the muscle chains in the body into optimal tension where the animal can feel free in that moment, safe, and begin to grow.

The dogs that I see in my practice are well cared for and safe. The lack of peace and comfort is from their past.

The simple sitting on the hands exercise can be helpful in bringing us back to the here and now so that our pets can have an experience of their actual here-and-now safety.

Sharing Life with Our Pets: Achieving Harmony Without Control

Most people seek out professional help for their animals when there is a problem. But it is actually advantageous to work on understanding the nature of an animal when things are going well.

When we choose to have a pet, we do so to add something to our lives, a natural loving energy that brings movement, love, and sometimes a feeling of contentment to our homes. There are many opportunities to understand nature more deeply, to be touched more deeply than is possible in our daily routine, and to feel a kind of companionship that requires no talking or thinking. In the process of living with and loving an animal, we have a chance to get to know who they really are and what we really are.

The true nature of an animal can be camouflaged by obsolete survival behaviors, for example, anxiety or aggression. Sometimes, the solution for these issues is love – giving what the animal needs when they need it and are receptive to it. As mentioned before, when an animal is launched from their subconscious into survival behavior, they are not receptive in that moment. Giving love – in the form of touch or a soothing, love-filled voice, giving food when they are hungry, maybe snuggling – can provide a new experience of loving safety. The accumulation of positive experiences results in trust and anchors the animal in the here and now.

Helping an animal forget the past happens in the here and now. Trying to compensate for their negative past usually fails. Supporting their here-and-now experiences by denying or not engaging in their neurotic behaviors provides the optimal circumstance for them to let go of their past. Trying to convince them of what they have or "bribe" them ultimately doesn't work. Here-and-now exercises, as discussed previously, are supportive. Giving a loving, neutral touch to the pelvis (hips, sacrum, pubic bone, wing bones) and balancing with an equal touch of the skull can give them a here-and-now experience. This is most effective after we humans sit on our hands or butt walk – anchoring ourselves first.

Space and the Relationship with the Self

Our pets depend on us to provide opportunities to experience satisfaction in their lives. Beyond our responsibilities to provide for them, we are most often (if not always) their primary relationship. They need us more than we need them, and yet there is an equality in the relationship.

They need circumstances to meet and experience themselves for themselves, to have an experience of themselves that is complete, unaffected by our influence. It's our responsibility to create this opportunity for them. For example, freedom to explore or freedom to be quiet, undisturbed when they need it. In other words, they need, at times, just to have their own space. Typically, in our culture, we care for them by becoming busy – with their diet, exercise, and their physical surroundings. Sometimes, they just need time to be together with us without an agenda. We can have blind spots for those times when they need safety from us, including safety in relationship to the outside world and emotional safety.

To have a natural relationship with us, they need to have a natural relationship with themselves. Sometimes, that means that they need a nest from us; sometimes, they need active, emotional exchange, and other times, they need boundaries.

It is important to feel what our pets need, as well as feeling ourselves in relationship to them and what our needs are. Sometimes, our needs are in conflict.

Pure perception can happen when we're in the here and now. As mentioned in a previous chapter, the exercise of sitting on our hands and relaxing our body so the weight of our trunk comes into our pelvis and into our hands, palms up, is a way to enhance our consciousness of the here and now. There are many other exercises that can support the growing of our consciousness and anchoring of our health and the health of our pets that are beyond the scope of this book. See resources at the end of this book.

Living in Peace with Our Pets

To be ourselves, relaxed in the here and now, is a foundation for health and happiness. The same is true for our pets. Training is a discipline to get animals to conform to our way of living and our societal standards while teaching necessary commands for their safety. When animals disturb our way of living, we work to control them so that the disturbing behavior is suppressed. Yet the underlying issues still remain.

Animals (and people) are restricted by their past, their early experiences as rapidly developing beings. They have a fear of being who they really are. As a result, they can become *too much inside* – isolated in their emotional experiences, shy, and reluctant. They don't bring their inner world (their emotions and preferences) to the outer world. They can also become overly reactive to the outside world and to each other. It is easy to recognize these behaviors in our pets. Cats can be withdrawn and disconnected, or they can be hostile when approached. Dogs can be aggressive, not necessarily dangerous, but aggressive in their approach to other dogs and people. These behaviors are adaptations to fear or developmental gaps and are not a reflection of their real personality. This fear is based on their subconscious awareness of their developmental gaps, a subconscious touch of pain or feelings of incompetence that translate to a feeling of unsafety. When an animal reacts in fear or aggression under safe circumstances, they are reacting out of a shock. This response is not a manifestation of their true nature.

So many of our pets come from difficult beginnings—abandoned, left without having the resources to fend for themselves, neglected, and/or abused. Many will be in life-threatening situations. Some will perceive that they are in a life-threatening situation when they are safe. Mistrust for life becomes anchored in their bodies and remains so, even when their here-and-now reality is perfect. These animals cannot see or feel what they actually have. They live in a chronic state of deep tension. Of course, when conditions are just right, they can find some relaxation and peace. But they can easily be triggered by the outside world and become anxious. When triggered, they lose their anchoring in the here and now and become anxious, afraid that something bad or even life-threatening will happen. This often occurs in the absence of any real threat. These triggers can include guests coming into the house, meeting other animals in the outside world (on the street, hiking trails, or dog parks), loud noises, etc.

Trust. Being reliable in reflecting our love and the safety of the circumstances gives the food that will grow and foster peace. And believing in ourselves and our love – and that what we provide – contributes to the healthy future of our pets. Our animals need us and our trust that they can and will learn, grow, and heal. We also need to trust ourselves, be patient, and not get hung up on failure or perceived failure. We need to allow ourselves to try new things without judging or blaming ourselves, other family members, or our pets when the results are not what we expect or desire. Natural growing is learning by making mistakes – stand up/fall down, two steps forward, one step back. This trust results in a beautiful togetherness. And peace.

Niek Brouw and the Foundation for Motor Development

Medicine has evolved from a discipline to an industry, an industry based on addressing health issues without patient participation and in a manner that does not reach the root of the problem. We have become dependent on the medical field to help us control or resolve our issues without really understanding what our issues are and what they have to do with us. Where do illnesses really come from? In my work as a veterinarian for 35 years, I realized that we do not have a deep and true understanding of the origin of many diseases.

The work of Niek Brouw illuminates and demonstrates the developmental origins of disease: genetic, fetal shock, birth trauma, and post-natal developmental shocks. By identifying and recognizing the shock and the developmental phase where it occurred, resolution can be achieved through exercises, manual techniques, and awareness. Through patient education and guidance, health can be restored, very often, without societal medical interventions.

Independence and self-responsibility are outcomes of Niek's method, along with continued growing and health.

The Foundation for Motor Development has been created to archive Niek's work which includes 40 years of recorded information to aid in the translation of his written works. The importance of Niek's work and his contribution to the world are beyond words for me. As human beings develop and evolve, I believe that Niek's work will be more universally recognized and embraced.

To learn more about Niek and the Foundation, go to niekbrouw.nl.

Afterword by Anne Lamott

I took my animals to Rob Erteman for 33 years for well-baby checks, well-elderly checks, and every age in between whenever they were sick or struggling. I first heard of him from friends with pets who loved him, so I took an old dog in for a consultation about her allergies, and it turned out that Rob was more than your ordinary vet: way more. Because of the research and experience of which you have been reading in this book, he was a kind of medical/psycho-spiritual practitioner who healed them in body, mind, and heart.

Let me tell you about a few of these precious animal friends.

My son and I had a house cat named Boo-Boots, who was suddenly old and infirm, and who definitely did not want to get into the cat carrier and go for rides. So Rob dropped by for coaching sessions. He felt strongly that she did not need medical intervention and taught us that while a big part of her was failing, 90% of her was just as comfortable and fabulous as ever. So my son and I tended to that 90%, talking to her, stroking her endlessly, and getting her to eat Rob's secret cat food suggestion, smoked oysters, in which we could also hide her medicines. He stopped by a few times near the end to consult with her in the mysterious way he communicates with animals, even cranky old cats – one time (and I am not making this up) during the Super Bowl. I mean, greater love hath no man.

Our darling dog Sadie, who had been the most perfect dog ever, like Jesus in a fur coat, got cancer at 10 years old. (My son told people she was a flat-coated retriever because he said it made us seem fancier, rather than the truth, which was that she was a mutt.) I could not bear for my son to lose her at such a vulnerable age (let alone darling vulnerable me!) or for Sadie to undergo the rigors of chemotherapy. I so hoped Rob could let Sam have one more season with Sadie, and make peace with her death. So after consulting with Sadie, and touching her in the slow and gentle way in which he gathers patient information, Rob came up with a cocktail of low-dose chemo, a steroid, and nutrients. She never once got sick from it, and lived another two years in good shape, loving life, looking forward to walks in the sunshine.

When it was her time to go, Rob again discussed how she did not need to make the trip to his office but rather just needed oral pain and anxiety meds. Sadie went under my bed for the last couple of days, so I gave her the medications and lay with her. Rob communicated that he wanted to check in on her breathing, and so I held the phone to her mouth and let him listen in. He said she was doing just great, that passing on is a lot like being in labor and that she was in a good place. I remember him saying, while I cried on the phone, "She has painkillers and you – we should ALL be so lucky."

My hardest dog was named Bodhi, a big lunk in a tuxedo, with enough pit bull mixed in with an even more difficult past so that he had shown some signs of aggression. Rob started seeing him on a regular basis, doing this medical/psycho-spiritual voodoo he does, and I'd watch the two of them with my mouth hanging open: Bodhi would start off growling and baring his teeth when Rob would first enter the exam room, and Rob would speak to him with respect and maybe a little bit of, "Oh, for Pete's sake, dude, it's just me." Bodhi would indicate an acceptable boundary where Rob could approach without risking attack, and Rob would switch to talking to both of us and let Bodhi experience simple interconnectedness. Then he would begin the psychomotoric work he does, touching Bodhi in certain spots, typically between the shoulder blades, which Bodhi would agree to, and then, within minutes, Bodhi would come lay at Rob's feet where they talked.

There have been so many cats and dogs he's tended to over the last 30 years – and even once my son's pet iguana, Barry. I could tell you how Rob worked with each of them – a kitten with feline leukemia named Cindy Lou Who and my old union cat Jeanie, who worked daily, tirelessly carrying my underpants out to the front steps. (I think she was flirting with the postman.) But I hope these specific stories have given you a sense of the man. The notes you have been reading are rare and extraordinary observations. I can't think of another book where you will find the insights of such a learned and profound veterinarian, animal lover and teacher, hands-on and practical and ethereal, who bridges the gap between humans and their pets, helping us help them, helping all of us experience the fun and rich love of our lives together.

Resources

niekbrouw.nl

vitaal.nl

juliefishmanhealingarts.com

juliewatershealingarts.caard.co

Tim and Jane Heintzelman Healing Arts ~ 415.454.2151

roberteman.com